simple

paper

* style *

simple
paper

❊ style ❊

DOROTHY WOOD

Photographs by Lucinda Symons

Watson-Guptill Publications/New York

Acknowledgments

Special thanks go to Lucinda Ganderton for designing and making the leaf blind, page 68
and the tulip collages for the screen, page 70; and to Purves & Purves, London, for the home accessories.

Suppliers

Kate's Paperie, 561 Broadway, New York, NY 10012, Tel: 212-941-9816.
For Lx or lens tissue paper: Bookmakers, 6001 60th Avenue, Suite 101, Riverdale, Maryland 20737,Tel: 301-459-3384.
The Paper Shed, March House, Tollerton, York, YO6 3EQ, England, Tel: 011-44-1347-838253.

Published by MQ Publications Ltd
254-258 Goswell Road, London EC1V 7EB

Copyright © MQ Publications Ltd, 1999
Text © Dorothy Wood 1999

First published in the United States in 1999
by Watson-Guptill Publications, a division of BPI Communications, Inc.
1515 Broadway, New York, NY 10036
Library of Congress Catalog Card Number: 98-61150
ISBN: 0-8230-4806-3

Series Editor: Ljiljana Ortolja-Baird
Editor: Alison Moss
Designer: Bet Ayer
Photographer: Lucinda Symons
Stylist: Fanny Ward

Printed in Italy

1 2 3 4 5 6 7 8 9 / 06 05 04 03 02 01 00 99

contents

introduction

✳

Modern style is often equated with minimalism, but living in surroundings as empty as possible is simple style without reality. First and foremost the design within a home should support and provide adequate space for the activities of busy family life. Discreet storage that complements and blends with other furnishings eliminates the chaos created by clutter; well-designed lighting and window arrangements enhance the mood of a room; and elegant knickknacks dotted about add the finishing touches to the ambience generated by the main decorations and furniture.

Author, Dorothy Wood addresses this reality, by suggesting projects with clean, simple designs, which also have practical appeal. Simplicity, however, is not to be confused with simplistic – it doesn't rule out precision, or proficiency in craft, but allows for the individuality that is inevitably a signature of homemade goods. And what better medium to work in. The natural quality of paper, and the ability to make use of old packaging and wrapping in some of these projects ties in perfectly with today's trends of reuse, rework, and recycle.

The beauty and quality of the various papers that I have used speak for themselves. The bold, sometimes earthy primary colors, and soft, subtle pastel shades, with minimal applied decoration create a timeless and tasteful appeal that will complement many styles and tastes.

All of us at some time or another have despaired at the growing pile of junk mail, the heap of newspapers and magazines on the coffee table or the towels strewn all over the bathroom floor. Although the home is basically a storehouse of belongings, sometimes the sheer bulk of our posses-

sions threatens to overwhelm us. The first chapter offers five stylish solutions to our most common storage problems: an elegant and copious storage bin, a stamp-covered box for important letters, shoe boxes with a fascinating cutwork design, a large storage box for toys or laundry and a designer magazine rack.

Once we have managed to solve all our storage problems, the surfaces will be free to display

the wonderful designs in the table top chapter: delicate lace paper bowls, a practical woven basket, a group of tall vases, some quirky coil pots and a collection of picture frames to display your most treasured photographs.

Whether it is a bright light directed onto a work surface or soft background lights for entertaining in the evening, different rooms in the house require versatile lighting solutions to accommodate their various functions. The third chapter includes an enchanting wall light, and a classic parchment shade that can be used with a lamp base or as a hanging light. There are also two eye-catching but totally different floor lights and a delightful tucked and stitched paper lampshade.

The fourth chapter concentrates on window dressing, providing five attractive but practical alternatives to curtains. There is a beautiful three-fold tulip screen, a charming shelf border, a unique faux window box, and two distinctly different blinds. One is a translucent flat blind with leaf motifs that makes a stylish alternative to lace curtains or frosted glass and the other is a concertina blind inspired by the wonderful colors and linear designs of the 1950s.

For some of the projects, I selected a variety of papers from around the world, such as exquisite lace papers from Japan for the delicate papier-mâché bowls, coarse handmade khadi paper from India for an unusual window screen and fragile plant tissue paper from Thailand to make a wall light. For others, I used recycled papers and materials to great effect: a selection of stamps decorate a letter chest, food wrappers make a wonderful design feature for a Japanese-style lampshade, and cardboard packaging is cut to shape for the laundry box and magazine rack. Strings and flat yarns are perhaps my favorite type of paper. They are extremely versatile and available in a wide range of wonderful col-

ors. They can be coiled, wrapped, woven, and even crocheted or knitted to create amazing textured surfaces.

Projects also try to make use of the inherent qualities of the chosen paper. For example, the fluting inside single-ply/wall cardboard is a major design feature on the magazine rack and plant fiber paper adds an extra dimension to the concertina wall light and leaf blind. I haven't used

many expensive papers, but have instead chosen to concentrate on getting the best possible results from the more basic bond/cartridge paper, sugar cane paper and cardboard. While researching, I found some unusual papers such as Lx/lens tissue/tissuetex, a fine tissue paper that behaves like fabric. This paper has wet strength that makes dyeing a simple process and its unique construction can also cope with machine stitching: both techniques were used to great effect on the tucked and stitched lampshade.

Before beginning any project think about where to find the necessary materials. Handmade paper and the more unusual products such as Lx/lens tissue/tissuetex, ribbon, string and flat yarn are often sold only in specialty stores in larger towns and cities but can also be ordered by mail. Rather than buying large sheets of cardboard from a packaging manufacturer ask friends and neighbors – or local liquor or grocery stores that may have extra packaging material. Another possibility is a framing shop where large sheets of mountboard are delivered in plain cardboard packaging.

The projects take differing amounts of time to complete. The Japanese bowls and coil pots may take only an hour or two, whereas the torn paper tulip panels for the screen require time and care.

If some of the techniques are new to you, try a small sample before starting on the actual item. A few projects require considerable sewing or crocheting skills. The projects have clear detailed instructions to make them accessible to all readers, but you can also be adventurous and try different types of paper or techniques for different projects. For example, decorate a storage box with the cutwork technique used for the shoe boxes or use the paper fasteners in the shelf border project to attach simple motifs to a lampshade or storage box. Once they are finished and varnished, paper items can be sponged clean from time to time; or use a hair dryer to remove the dust.

Paper is such an integral part of our lives and is used in so many different ways that we often tend to take it for granted. This book maximizes paper's potential and acts as a showcase for 20 different projects designed to be attractive yet safe, durable and ideal for their intended purpose. The range of basic techniques such as cutting, folding and tearing, combined with the more challenging skills of weaving and crochet will prove stimulating, and the finished results will bring great satisfaction.

INTRODUCTION

storage

▶ 11

Scattered around a typical home are an assortment of possessions that could benefit from a better storage system. Whatever the need, storage items must blend in with the surroundings and suit the purpose for which they were designed. I have chosen a selection of items likely to be used in every home, such as a shoe box strong enough to protect a pair of shoes; a letter chest just the right size for paper and pens; and a laundry box large enough to contain the family wash.

There are many cheap cardboard accessories available in stores, and a few simple ideas can transform a basic storage box or three-drawer chest. At first glance, the design on the shoe boxes looks like a random collection of intricate cut motifs, but the design began as a series of "S" shapes

drawn out to fit the lid. Use these scarves, shoes or stockings. The strength of the box, so several can

The set of drawers takes on a with old stamps and finished with age bin uses a similar technique. pasted onto dense cardboard and the lets with a soft green cord.

It often seems to be the case

decorative painted boxes to store cutwork will not greatly affect the be stacked with confidence.

completely new look when covered some stylish string handles. The stor- Sections of a flat map have been panels are tied together through eye-

that the laundry basket is full to overflowing or the children's toys are strewn all over the living room. The large brightly colored laun-

dry or toy box is the ideal solution to both these situations. Long strips of cardboard, dyed or painted in bold colors, are woven together and finished with an attractive stitched edging.

The magazine rack certainly looks like a designer item and will complement any living area. The long flowing lines are easy to cut out with a craft knife and the only precaution is to make sure you have enough cardboard to cut.

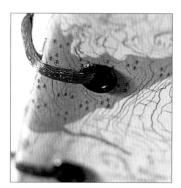

STORAGE

storage
bin
✳

This large bin, covered with sections from a map, will hold all of the oversized odds and ends of a busy family room or study. Look for an attractive map in the travel section of your local book store, then ask for it as a "flat" map. This type of map comes rolled, rather than folded and so has no ugly crease lines.

The sides of the bin are made from a dense cardboard that can be cut using a large craft knife. Use a rotating hole punch to make holes in the cardboard that are just big enough for the eyelet to fit. Because the board is so dense, clear the punch after making each hole to make the task easier.

▶ 15

The decorative eyelets holding the sides together echo the spiral binding on a notebook.

magazine
rack

✳

The fluting inside the cardboard is a major design feature of this magazine rack. It may seem daunting to cut out so many pieces of cardboard but the soft flowing lines make the task easy. The template has been made specifically so that each shape butts against the previous one, thereby reducing the required cutting by almost half. The pieces are marked out in groups of five, with the fluting going across the width to produce an attractive zigzag design down the side. Replace the blade on the craft knife regularly to prevent the fluting from tearing. The pieces are stuck together with a strong wood glue to create a very sturdy structure that will last for years. As an alternative, the plain cardboard structure could be covered in papier-mâché or spray painted.

Overleaf: Detail of the magazine rack shows how the fluting inside the cardboard is used to great effect.

laundry
box

✳

Wide strips of corrugated cardboard have been woven together to make this large box that would be suitable for the storage of laundry or toys. The large sheets of single-ply/wall cardboard were originally used to box unassembled furniture but they can be obtained directly from packaging companies. The brown cardboard dulls the bright wood dyes used to stain the strips, to produce attractive rich deep colors.

Use three coordinating colors of paper string or fine cord to stitch an attractive border around the top edge of the box. The overlapping stitch pattern produces a strong eye-catching edge. You could also make a shallow lid for the box. Cut the strips slightly wider so that the lid fits over the top, and finish the edge of the lid with the same decorative stitching.

▶ 21

The bright colors in the stitching pick up the strong colors of the box.

shoe
boxes

✳

All of us have several pairs of special shoes that are kept in their original boxes until the proper occasion. These boxes are not particularly strong and if piled in the bottom of the closet may be doing more harm than good. Strong basic storage boxes can be purchased from discount stores and transformed with this stunning cutwork design. The pattern is based on the letter "S" with the intricate spaces between each letter cut away. Use a very sharp craft knife with a long narrow blade to cut around the deep curves. It is essential that the blade is very sharp to cut through such strong cardboard: I found that I needed to use at least two blades for each box.

The flowing cutwork design, allows you to recognize the contents without having to lift the lid.

letter
chest

✳

There are many simple, well-designed cardboard accessories in stores today that are generally sold unassembled and are folded to make durable, practical storage items for the home. This chest was made from a set of drawers of single-ply/wall corrugated cardboard, assembled by folding the corrugated side to the inside to create a smooth surface for the stamps. All the stamps I used were rejects from stamp collections. They have little surface value but are ideal for decorating the chest of drawers.

The elaborate handles are made from a deep blue paper string that is wrapped around a strong inner core. Paint the chest with a matte acrylic varnish to protect the surface.

▶ 25

The bright coiled string handles make a stunning finish to the stamp-covered set of drawers.

table top

A ll over the home there are surfaces that could be brightened up with a stylish accessory such as a bowl, basket or vase. Sometimes care must be taken to protect the surface from being scratched, but with all these projects made from paper, surface damage will be minimal because of the type of paper and techniques used.

Although bowls and vases are primarily designed for storage, they may look even better if left empty. The elegant white lace paper bowls described on page 40 look so delicate against a beautiful wood table, it would be a shame to put anything in them!

I have devised a variety of shapes and sizes of containers to suit most situations in the home. The tall vases are designed to sit on a sideboard or table. Placed near a wall, they can be used to display a few dramatic stems of dried foliage or flowers, or even some of the striking artificial flowers you can buy these days. The woven basket is more utilitarian. It is the ideal size and shape in which to keep your sewing kit or knitting. The deep sides of the basket will keep the yarn from rolling all over the floor.

Depending on the size you choose to make, the distinctive coil pots can have many uses. The large pots are ideal as planters provided you use a plastic saucer underneath the plant pot to avoid water damage. A flatter version could be used to hold fruit or nuts and small pots are just what you need in the bathroom to hold cotton balls or soap.

Everyone has a selection of favorite photographs or pretty cards to display, but often in a disparate range of frames. On pages 34–37, an assorted collection of frames have been given separate treatments with tissue and flat paper yarn in coordinating colors that work together to create a sense of unity yet complement the different images.

TABLE TOP

tall
vases

❋

The beauty of these vases lies in the simplicity of their construction and the ease with which the colors can be changed to suit their surroundings. The shape is built from five simple pieces of cardboard taped together. This basic structure is covered with an attractive crinkled paper ribbon and painted.

Paper ribbon is sold in flower shops and gift wrap departments. Sometimes it is sold twisted into a thick rope that needs to be unraveled before use. As the ribbon is quite wide, it is cut into narrow strips to make it easier to wrap around the shell. Acrylic paint is ideal for painting the vase as the color is very even and it dries quickly. Spray the vase with a coat of clear acrylic varnish to make it look like ceramic.

The finished vases can be given extra stability by filling them halfway with silver sand.

c o i l
p o t s

✳

Coiling is a simple technique used to make pots from clay. Long sausage shapes are coiled and the ribbed surface smoothed down to create a flat-sided pot. In this case corrugated paper has been coiled and then covered in tissue paper to smooth out the ridges.

Corrugated paper and tissue paper are basic packaging materials used to send prints and posters by mail. Why not try recycling them to make some of these unusual pots in different shapes and sizes. The corrugated paper is cut into narrow strips and held in place with PVA glue. The attractive pattern is achieved by overlapping pieces of torn white tissue that are pasted in layers over the surface to produce pots that are remarkably strong.

These sturdy little pots are also ideal for holding plants, fruit, nuts and candy.

picture
frames

❄

When we decide to redecorate, some of the old accessories no longer seem to fit the new colors and furnishings. Items such as picture frames can be revamped rather than consigned to the closet or garbage can. In fact, a whole collection of similar frames can be color-coordinated to suit the new decor using simple paper products.

Paper yarn is made in a similar way to paper string but it is flat rather than round. The yarn has an attractive irregular appearance because it is not tightly twisted. Four soft colors were chosen to make the woven frame and two individual colors were selected to make the coiled and wrapped frames. The latter of these uses a narrow flat ribbon. If this is difficult to find, you could cut strips of tissue paper instead.

woven
basket

✳

Wicker storage baskets sparked the idea for this traditional-style basket made entirely from paper. Fine paper string has been woven over a single-ply/wall cardboard structure. The slits in the cardboard need to be wide enough to take the paper string without distorting the shape of the basket. Cut out part of the template and weave a small sample piece to check the width of slot required for your particular string. The slot should remain the same width once woven. If it gets narrower, the slot is too wide whereas if the gap widens the slot is too narrow. The subtle neutral colors have been chosen to complement the brown color of the cardboard but the cardboard could be painted to match brighter colors if preferred.

▶ 39

lace paper
bowls

❋

Papier-mâché is normally made with newspaper, but any type of paper can be used. The exquisite lace paper from Japan used to make these bowls is very delicate but becomes quite stiff when brushed with PVA glue. It is produced in a variety of patterns that can be used to great effect to create some unusual and pretty edges to the bowls. Paper with a pattern of tiny squares tears very raggedly and produces a rough feathered edge, whereas paper with a circular pattern forms an attractive scallop edge. Tear any solid areas out of the paper before pasting to keep the bowls light and airy. These papier-mâché bowls can be made using any plastic container that is wider at the top than around the base. Make a range of them in various shapes and sizes with different textures.

40 ◄

A layer of pressed leaves or petals adds splashes of color to the brilliant white. Although they look stunning empty the bowls can hold items such as cotton balls, potpourri, nuts and candy.

lighting

▶ 43

W ell-composed lighting can greatly enhance the atmosphere of a room, and strategically placed lamps can alter its appearance quite considerably. A floor lamp draws your attention naturally downwards, whereas a ceiling or wall light has the opposite effect. Although lamps are designed to be illuminated, most of the time they are switched off and as a result have to be just as attractive when viewed during the daylight hours. The concertina wall light, for example, has a pretty gilt and ochre yellow motif that blends with the speckled background but this becomes a much stronger image when lit and the gilt leaves and tobacco flecks are thrown into relief.

A more subtle change occurs with the tucked and stitched lampshade. The translucent nature of the tissue paper layers becomes much more evident when the lamp is lit and the differing densities of paper allow varying amounts of light to pass through. This project requires basic sewing skills.

Many of today's furnishings are quite low and floor lamps help to achieve a balance by drawing attention to a pretty rug or changing the mood of an empty corner. The Japanese lamp and crochet lamp base are both designed to stand on their own. The wonderful crisp texture of the crochet on the latter contrasts wonderfully with the soft wool of a carpet or rug.

Parchment and natural raffia are a classic combination that has been popular for several years. For the parchment lampshade, small V-shaped notches are cut in the parchment to create a simple

pattern that is equally pleasing whether lit or not.

Finally, a word or two on safety. There is a fire risk when using paper to make lampshades. Follow the manufacturer's guidelines to ascertain the maximum wattage of the bulb for each type of lamp and ensure the bulb doesn't touch the shade when assembled. As an additional precaution, treat the paper components of all lamps with a fire retardant spray.

LIGHTING

Japanese
lamp

✳

The simple lines of Japanese furnishings were the inspiration for the design of this classic lamp. The frame is built from metal rods which have been welded together and then sprayed with matte black paint. The frame is not completely square, but slopes out gently to give a more balanced appearance.

The paper used to cover the frame is a traditional paper from Japan called shoji. This is a strong tissue paper with a wonderful soft texture generally used for calligraphy. It is fairly translucent and allows the light to filter through the different layers of paper that decorate the surface of the shade. Look for attractive packaging made from unusual paper in Japanese supermarkets or delicatessens and select some Japanese text that can be enlarged and copied onto the shoji paper.

The essence of Japanese interior style is recaptured in the elegant paperwork design.

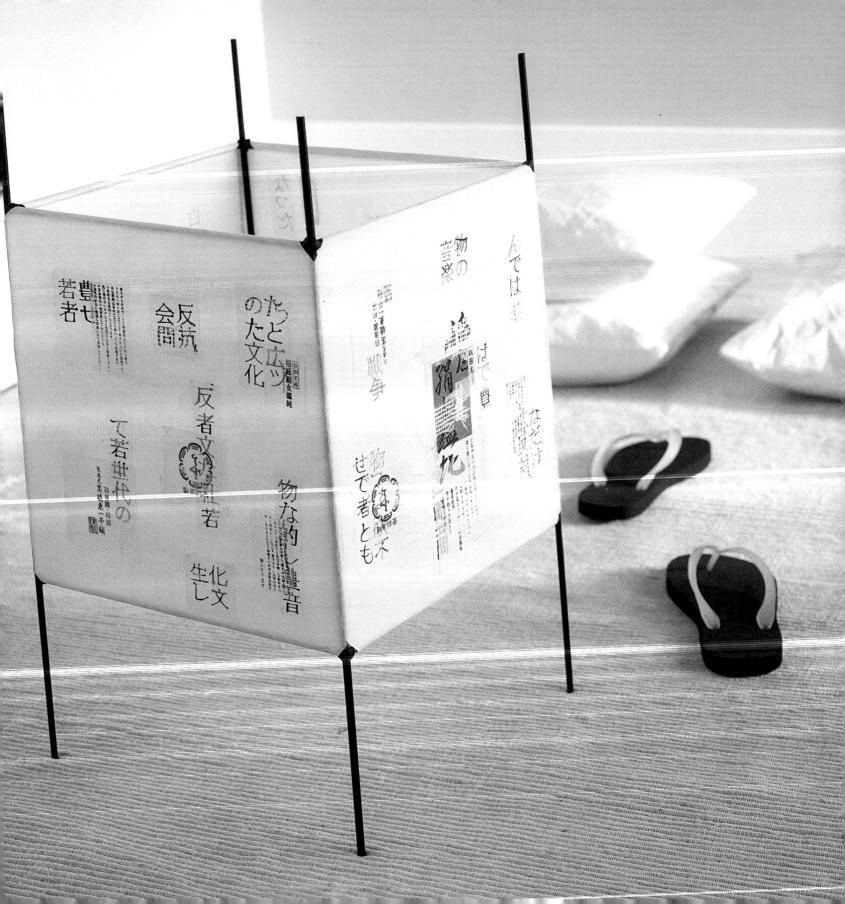

SIMPLE PAPER STYLE

concertina
wall light

❋

Wall lights have a dual purpose. During the day they are purely decorative – a sculptural wall decoration – but when lit they completely change appearance, casting a soft light over the walls. They are ideal for dining rooms, allowing you to have candles on the table but still have enough light to see what you are eating.

The gold leaf and ochre yellow stencil motifs blend with the unusual tobacco paper to produce a most attractive wall decoration. In the evening the light inside throws the leaf motif and brown flecks of tobacco into relief to create a different but similarly pleasing effect. Use a semicircular wall light fitting as a base for the concertina and fit a bulb as recommended by the manufacturer.

▶ 49

Overleaf: Detail showing the stencilled leaf motifs that decorate the concertina wall light.

tucked and stitched lampshade

✳

I have always loved the way different shades of color can be built up using layers of tissue paper and so I designed a lampshade that would make use of this effect. The pale green shade was not available in ordinary tissue paper and attempts to dye tissue at home ended in failure. Success came with Lx/lens tissue/tissuetex, a new tissue paper that can be handled like fabric. It has wet strength, which means that it dyes easily and can also be stitched without tearing. Buy a blue/green cold water dye or mix turquoise, chrome yellow and cerise Procion "m" dyes to produce this crisp shade of mint green. The depth of color can be altered by the addition of water but remember that the color becomes lighter once dry.

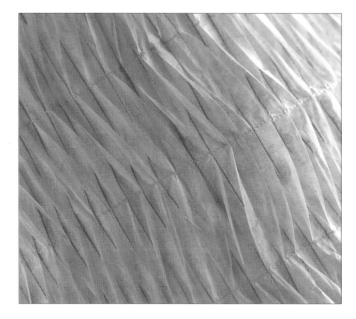

The rippling motion of the tucks has a calming effect, and softens the light passing through.

crocheted
lamp base

✳

Paper string has been available for years for gift wrapping. It is generally sold in small skeins with two or three strands twisted together to make a fairly thick cord. Although it is possible to unravel the strands for this project, it would be easier to look for skeins of single ply. Choose four shades of one color that will blend together to produce an ombré effect. The lampshade is worked in double crochet in one long panel and requires considerable crocheting skills. To keep the sides of the panel straight, crochet the same number of stitches in each row and work the last stitch into the turning chain. This is an extra single chain worked to bring the yarn up to the level of the next row of double crochet. The coiled wire feet lift the lamp off the floor or table and allow the cord to fall away at the back of the lamp.

▶ 55

The colors of the crocheted shade can be varied to blend with a favorite rug.

p a r c h m e n t
s h a d e

✳

The mellow shades of parchment paper and natural
raffia work so well together that there is no need here
for additional embellishments. Traditional parchment
is made from sheepskin but a similar parchment
paper is more readily available. This type of paper is
ideal for the delicate cutwork needed for this project,
because its stiffness allows the shade to be self sup-
porting between the two lampshade rings with no
ugly uprights to spoil the effect. The simple design
can be adapted to fit any shape or size of shade. Use
your old shade as a template and mark the position of
the tiny notches carefully. Because the parchment is
translucent you can cut the notches directly using the
template underneath as a guide. A sharp craft knife is
essential to prevent the parchment from tearing.

The basic V-shape design
creates illusory lines and circles.

window dressing

F or the past few years, windows have been considered unfinished without large swags or flounces on either side of the curtains. With current trends now moving towards a more simple approach, I wanted to find alternative ways to dress a window.

Windows exist primarily for letting light and air into the house and are usually completely covered at night to keep the heat in and provide privacy. Sometimes windows also need screening through the day to provide some seclusion from the outside world. A translucent flat blind is a wonderful, stylish alternative to frosted glass or lace blinds. Cut-paper leaf motifs, sand-

wiched between two layers of tissue paper, make a simple paper blind that is left down all day allowing the light to pass through while decorating the window with a pretty pattern.

The concertina blind serves a slightly different purpose, blocking out all light if kept down. I used an opaque thick bond paper to make this blind because it was easy to fold and decorate.

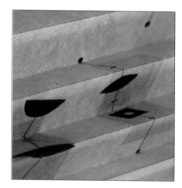

If a small window has an unattractive view, a series of shelves can be built into the window recess to display a collection of crockery or knick-knacks. The pretty bird motif shelf border would help to fill in the gaps and make an effective window screen.

The dentil border on the window box is another type of screen that can be used to hide an assortment of plant pots along the kitchen window sill. Cut out of thick cardboard, the border is glued onto a panel of medium-density fiberboard (MDF) and held in position by strips of wood at either side.

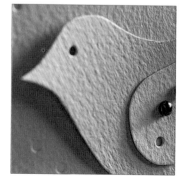

The final project in this section is the most substantial window treatment. The three-fold screen stands about shoulder height and is decorated with nine panels of torn paper collage. This type of screen would be ideal in the bedroom of a Victorian-style house with tall sash windows.

WINDOW DRESSING

concertina
blind

❉

After four decades, the color schemes and designs of the 1950s are popular once again. The design on this blind is deceptively simple. The wonderful, wacky motifs are worked in stages using basic ink stamps and a black pen. The blind is made from a thick textured bond paper that is made in France. This paper was chosen because it folds very neatly and is available in a wide range of colors.

All papers tend to crinkle to some extent when wet, so although the pale green panels look painted, they are actually stuck on. I used Lx/lens tissue/tissuetex, a strong tissue paper, that was dyed pale green and spray mounted to the background. Stamps were cut to make the red wine colored shapes and the motifs were completed with a fine black pen.

▶ 63

The striking motifs are created with abstract ink stamps linked with a black line.

window
box

※

Plants always make an attractive feature along a window sill but the odd collection of pots in different shapes and sizes can make them look unkempt. Plant pots can be hidden neatly behind the solid area of this imitation window box, with the ornate dentil work adding an attractive decorative finish along the window ledge.

Any type of stiff card that can be cut easily is suitable for this project. I have used khadi cotton rag paper from India that has an attractive surface texture. Although it is quite thick the paper is soft enough to cut with scissors. Khadi paper is available in a range of different colors and can be painted. If the paper buckles slightly, leave it under a flat board and heavy weight overnight to smooth it out again.

The repetitive flower pot border gives a neat and ordered finish to the window box.

shelf
border

✳

Plain wood shelves can be transformed with the addition of these pretty shelf borders. Or you could make an attractive feature in a small window by fitting shelves at intervals across the window recess and adding a shelf border to each one. The decorated shelves could be used to display a collection of jugs or tinware, affording considerable privacy or disguising an unattractive view.

The material used for the border is an artist's watercolor paper, called Bockingford, which is generally sold in white but is available in several pale pastel colors such as the gray and cream used here. The delicacy of this design is achieved using an adjustable hole punch to make small holes. The paper fasteners are also much smaller than normal but can be ordered from your local stationer.

▶ 67

leaf
blind

❖

This beautiful translucent blind acts rather like frosted glass, letting in the light while providing a permanent screen across the window. The blind is designed to give the appearance of leaves falling outside the window. The paper used to make the blind is Lx/lens tissue/tissuetex, a new strong type of tissue paper. Two layers are sandwiched together with spray adhesive, trapping the stylized cut leaves between them. Look for unusual plant fiber papers for the leaves. These are fine textured papers with pieces of plant material or flowers embedded in them which are revealed when the blind has light behind it. Sew two or three panels together to make a larger blind and hang using curtain clips or stitch onto a decorative pole.

The textured papers used for the leaf shapes add depth and dimension to the blind.

tulip
screen

✳

Tulips last for such a short time in their full blown shape, that I decided to capture their beauty to be enjoyed all year round. The tulip panels were designed from a series of photographs taken in the spring. Experiments using several types of paper revealed that basic sugar cane paper was the easiest to tear into the intricate shapes. Choose a selection of soft colors that work well together to create the flowers. The background panels can be cut from more interesting textured or handmade paper. Choose three paler shades to complement the sugar cane paper tulips.

The plain three-fold screen is painted in a dull ivory and distressed to show the original sage green around the edges. The panels are pasted on using diluted PVA glue and the entire screen can be sprayed with matte varnish to protect the paper collage.

Each tulip shape is built up from pieces of torn sugar cane paper in a range of sumptuous colors.

putting it together

All the projects in *Simple Paper Style* require cutting in one way or another. On occasion a pair of scissors is sufficient, but a craft knife is generally a better tool. For the storage bin and woven laundry box, use a heavy weight craft knife. Finer projects such as the parchment lampshade and shelf border will be easier to do with a pencil-style craft knife that can accommodate a variety of blade shapes. Choose a long pointed blade for cutting the intricate curved shapes on the shoe boxes.

Keep a good supply of blades for your craft knife. Blades become blunt quite quickly producing a torn, ragged edge, and are then dangerous because the extra effort required can cause the blade to slip.

It is also worth investing in a cutting mat and metal edge ruler. Other useful items include a

rotating hole punch for the storage bin, shelf border and concertina projects, and a rubber paint roller for the wall light, storage bin and tulip screen. You will need a crochet hook and basic sewing equipment for other projects.

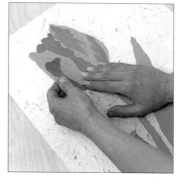

Most projects need to be glued. PVA is a versatile glue that can be diluted to different strengths. As a rough guide, use it straight for cardboard, watered down 50/50 for paper and with five times as much water for papier-mâché. Wood glue is an extra strong PVA ideal for sticking cardboard.

PVA contains a high proportion of water that causes some papers to expand and buckle. I found that parchment was particularly susceptible to moisture. In this case spray adhesive is better. Some brands allow the paper to be repositioned a few times if necessary but spraying both surfaces can produce an immediate strong bond. Double-sided tape also bonds instantly. Use small pieces to position the paper on the Japanese frame before gluing and for sticking paper to the frames. Fast drying all-purpose glue is ideal if the paper can't be held in position for long. Remember to always experiment first on a small sample of your chosen material.

PUTTING IT TOGETHER

storage bin

✳

Flat maps can be ordered from major book stores. Choose a portrait shape map that can be cut in four to cover the sides of the bin.

MATERIALS
- ◆ *Dense cardboard/millboard* ¹/₁₆*in/2mm thick*
- ◆ *Large flat map*
- ◆ *Soft green paper*
- ◆ *Soft green cord*
- ◆ *Spray adhesive*
- ◆ *Eyelet punch and eyelets*
- ◆ *Eyelet tool*
- ◆ *Pencil*
- ◆ *Scotch tape/Sellotape*

1 Decide on the size of the bin. Cut four rectangular panels from the board and a square with sides slightly narrower than the short sides of the rectangles.

2 Draw a line ½in/1cm in from the edge down the long sides of the rectangles and along one short side and around each side of the square base.

3 Punch holes every 1¼in/3cm along the pencil lines. Make the holes just large enough to fit the eyelets.

4 Cut the map into four rectangles and trim to ⅝in/1.5cm larger than the board. Spray adhesive on the back side of the map and on the board. Position the map on the board and smooth out any air bubbles with your hand.

5 Turn the board over onto the backside. Trim across the corners of the map, spray the edge of the board with spray adhesive, and then fold the map over and stick in place (*see picture next column*).

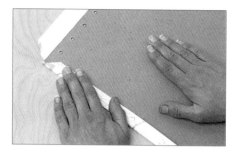

6 Cut the green paper to fit on the back side overlapping the raw edges of the map. Spray the paper with adhesive and stick down.

7 Punch the holes through from the right side with a pencil. Push eyelets into the holes and use an eyelet tool to secure.

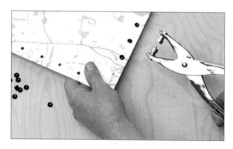

8 Wrap a piece of Scotch tape/Sellotape around the end of the green cord. Feed the cord through the holes to "sew" two panels together. Leave enough cord to join along one bottom edge and cut off.

9 Wrap a second piece of cord with tape and sew the third and fourth panels in position. Use the excess cord to attach one side of the base at a time. Tie the cord off neatly on the inside.

magazine rack

✳

This magazine rack is surprisingly solid when assembled. Make sure you have plenty of single-ply/wall corrugated cardboard before starting the project.

MATERIALS

- ◆ *Single-ply/wall corrugated cardboard*
- ◆ *Thick card stock*
- ◆ *Pencil*
- ◆ *Craft knife*
- ◆ *Cutting mat*
- ◆ *PVA glue*

1 Enlarge and transfer the template on to thick card stock and cut out the two pieces. Mark the grain line across the center of the large card stock template.

2 Lay the template on the corrugated cardboard matching the grain line to the corrugated flutes in the cardboard. Draw around the template then move it across and down slightly so that the curve of the template butts against the previous shape. Draw around the template again (*see picture next column*).

3 Continue moving the template and drawing around it until you have five shapes, then begin again in another area of cardboard. This ensures the corrugated pattern inside the cardboard looks the same on each piece.

4 Cut around the shapes with a craft knife. You will need 68 large pieces altogether.

5 Brush PVA glue over one side of the large pieces and stick 12 bundles of five pieces. Stick the last eight pieces in two bundles of four for the end sections.

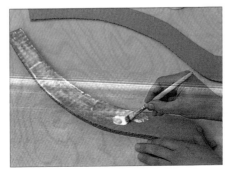

6 Cut and paste 68 pieces in the same way with the smaller template.

7 Tape the large and small templates together again to make an inverted Y-shape. Use this to draw and cut two shapes from the cardboard.

8 Paste one of the Y-shapes and stick a bundle of four small and four large pieces to it to make an end section.

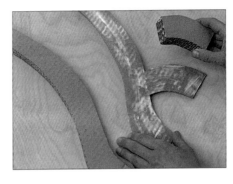

9 Position a large bundle over this to form a wish bone shape and mark the outline. Paste within the lines and stick in place. Stick a small bundle beside this to complete the next layer.

10 Continue building the layers until there are seven bundles on each side. Finish with two bundles of four and paste on the Y-shaped end piece.

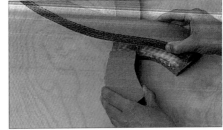

laundry box

✳

Large sheets of corrugated cardboard can be obtained from packaging companies or ask at local stores such as framers who have mountboard delivered in large flat packs of plain cardboard.

MATERIALS
- ◆ *Single-ply/wall corrugated cardboard*
- ◆ *Wood dyes – red, orange, yellow, green, blue, purple*
- ◆ *Paper string – yellow, red, orange*
- ◆ *Paintbrush*
- ◆ *Craft knife and cutting mat*
- ◆ *PVA glue*
- ◆ *Clothespins/pegs*
- ◆ *Large-eye needle*
- ◆ *Matte acrylic varnish*

1 Cut twelve 3in/7.5cm wide strips from the cardboard with the grain running lengthwise. Cut another twelve strips with the grain running across the width. Cut all the strips at least 56in/142cm long.

2 Dye two of each type with a different color of wood dye. Work in a well-ventilated room. Paint dye on both sides of the strips and allow to dry.

3 To form the base of the box, weave six strips (alternating colors) in one direction and six perpendicular so that the long ends stick out from a central woven panel. Use the strips with the grain running lengthwise if you want the grain on the finished box to run in the same direction.

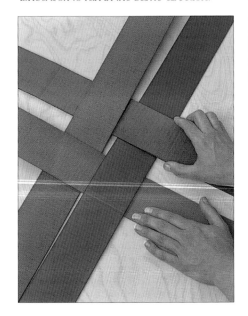

4 Crease the strips at the edge of the panel and fold upright to begin to form the box shape.

5 Using the remaining strips, weave one strip of cardboard through the uprights, crease to go round the corner and work in the next side. Use clothespins/pegs to temporarily support the uprights. Once the second and third strips are in place the box will have a little more stability (*see picture next column*).

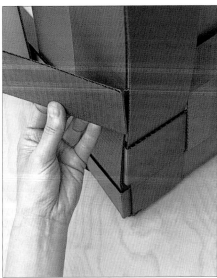

6 Keeping the same color sequence, weave a strip through the uprights and around the opposite corner.

7 Join the strips together on the inside by trimming the ends to overlap by 3in/7.5cm and gluing together. If the cardboard is bulky, tear away one layer from the end of each strip before gluing.

8 Continue weaving the sides in the same way until there are six horizontal strips. Fold the upright strips over the top horizontal strips. Alternate strips will fold to the inside and outside edges. Cut each flap to ¾in/2cm.

9 With the large needle, punch two holes 1in/2.5cm apart in the middle of each upright just under the top flap. Oversew the yellow paper string diagonally around the top edge, going into each hole and the gap between the uprights. Sew in the end of the string through the cardboard and snip off (*see picture next column*).

10 Oversew the red paper string diagonally in the opposite direction.

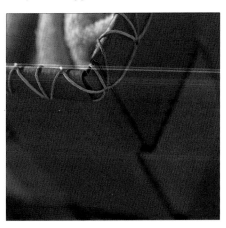

11 Stitch the orange thread vertically and then make a running stitch along to the next hole. Work vertical stitches over each hole. Complete the stitch pattern by working orange running stitch to fill in the gaps in the stitching.

12 Paint the entire box with two coats of matte acrylic varnish.

shoe boxes

✳

Pencil-style craft knives can take a wide range of different blades. Choose a long pointed blade to cut round the curves. Replace the blade frequently.

MATERIALS
- ◆ *Sturdy cardboard shoesize boxes*
- ◆ *Letter-size/A4 white paper*
- ◆ *Carbon paper*
- ◆ *Pencil*
- ◆ *Craft knife*
- ◆ *Tracing paper*
- ◆ *Flat acrylic milk-white paint/Repro-duction flat paint*
- ◆ *Paintbrush*
- ◆ *Round file*
- ◆ *Acrylic varnish*

1 Enlarge the template to fit the lid of the box allowing a 1in/2.5cm border all the way around.

2 Lay the carbon paper, carbon side down, on the lid and position the template on top.

3 Holding the template firmly, draw around the lines of the design to transfer it onto the box lid.

4 Score along the lines using a craft knife. Use a long narrow blade to follow the curves of the pattern.

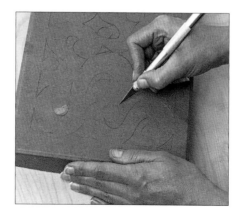

5 Cut into the cardboard along the score lines. Use the tip of the craft knife to go smoothly around the curves (*see picture next column*).

6 Cut pieces of tracing paper to the same size as the side and end panels of the box. Draw in a 1in/2.5cm border on each piece. Position them over the main template and trace off a design. Transfer the lines to the side and end panels. Score and cut as before.

7 Paint the shoe box inside and out. Once dry, file the cut shapes to smooth any ragged edges. Paint the box with a second coat of paint.

8 Paint the box inside and out with a coat of acrylic varnish.

letter chest

✳

Specialist stamp shops sell large packs of mixed stamps that are suitable for découpage. Choose a pack of older stamps with soft muted shades rather than modern glazed stamps.

MATERIALS

- ◆ *Corrugated cardboard three-drawer chest*
- ◆ *Brown gummed tape*
- ◆ *Stamps*
- ◆ *Wallpaper paste*
- ◆ *Hole punch*
- ◆ *Thick paper string*
- ◆ *Blue medium paper string*
- ◆ *Scissors*
- ◆ *All purpose glue*
- ◆ *Pen*
- ◆ *Matte acrylic varnish*

1 Make up the cardboard chest with the smooth side to the outside. Cover the slots and corner folds with the tape so that the surface is completely smooth (*see picture next column*).

2 Begin to paste stamps to the front of the drawers. Dot the stamps all over the surface before beginning to fill in the gaps. This prevents the stamps spiraling out from the one point and gives a more random appearance.

3 Stick stamps over the edges and about 2in/5cm down the sides.

4 Cover the sides and dividing panels of the chest in the same way sticking the stamps about 2in/5cm inside.

5 Mark the position of the handles on the front of the drawers. Punch a hole at each marking. Tie a knot in the end of the paper string and thread through one hole from the inside. Thread back through the other hole and tie it on the inside. Trim off the excess string.

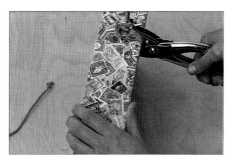

6 Spread glue around one hole on the front of a drawer. Working from the middle of the handle towards the drawer, wrap the blue string tightly around the handle, catching in the end as you go, and coil it around the hole to the required size. Trim the end. Unravel the string slightly, tuck the threads under the coil and hold until the glue dries.

7 Complete the other side of the handle in the same way. Repeat for each drawer. Paint the chest with matte varnish.

tall vases

✳

Paper ribbon is available in a variety of colors from florist supply stores. It is often used to make large bows for floral arrangements and bouquets.

MATERIALS
- Corrugated cardboard
- Paper ribbon
- Craft knife
- Cutting mat
- Brown gummed tape
- Scissors
- Pencil
- Acrylic paint
- Spray acrylic glaze
- Silver sand

1 Enlarge the template to 11in/28cm tall. Draw around the template and cut two pieces of corrugated cardboard. Cut two strips for the sides measuring 11 x 1³⁄₄in/28 x 4.5cm and a piece to fit across the base of the vase (*see picture next column*).

2 Stick the pieces of corrugated cardboard together with strips of brown gummed tape. Make sure that the bottom of the vase is level before attaching the base piece.

3 Open out the paper ribbon and cut in half lengthwise. Cut each half in two again to make four long narrow strips of crinkled paper.

4 Starting from the base, cover part of the vase with PVA glue and wrap ribbon strips around it, overlapping as you go. Keep the paper quite tight but encourage the natural crinkles on the ribbon to form (*see picture next column*).

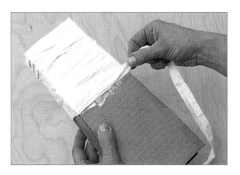

5 Continue pasting the vase and sticking down the ribbon. Add extra lengths as required. Tuck the end of each ribbon under the previous layer.

6 Allow the edge of the last layer of ribbon to stick up about ¹⁄₂in/1cm above the vase. Smooth out the top part of the ribbon and snip into each corner. Glue the overlap down on to the inside edge.

7 Paint the vase with two coats of acrylic paint. Allow to dry and spray with a clear acrylic glaze.

8 Fill the vase part way with silver sand to weight the bottom.

coil pots

✳

Corrugated paper is a cheap packaging material available in rolls from stationers. It has a single flat surface that makes it easy to coil.

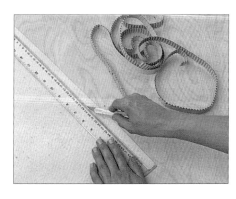

2 Coil the corrugated paper, brushing PVA glue on the rib side of the strips as you go. Keep the coil flat to form the base of the pot.

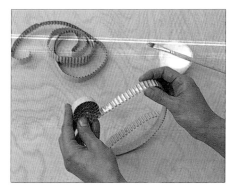

3 Allow the base to dry flat. Brush glue around the top half of the edge of the base and wrap a new strip of corrugated paper around sticking it down onto the glued section smooth side out.

4 Continue gluing and wrapping strips to form the bowl. The shape can be adjusted by overlapping the coils to different depths (*see picture next column*).

5 Once the coiled pot is the desired shape and size, level out the top by gluing the corrugated strip down onto the last full circle.

6 Pull the last ½in/1cm of the corrugated "flutes" away from the paper backing and glue down flat. Stick the backing over the rim to give a smooth finish.

7 Tear pieces of white tissue paper into long ragged shapes. Use diluted PVA glue to paste the tissue paper over the entire surface of the pot.

8 Once dry, paint with two coats of matte acrylic varnish.

MATERIALS
- Corrugated paper
- White tissue
- Craft knife and ruler
- Cutting mat
- PVA glue
- Matte acrylic varnish

1 Cut several ½in/1cm wide strips from the corrugated paper with the ribs running across the width (*see picture next column*).

picture frames

In this project, three basic picture frames have been covered using different techniques in a range of harmonizing colors of tissue paper and flat paper yarn to give them a new lease on life.

Flat paper yarn, used to decorate the two frames in the foreground of the photograph below, is a versatile material that can be coiled, woven, knitted or even crocheted to produce all sorts of attractive effects.

MATERIALS

- ◆ *Basic picture frames*
- ◆ *Flat paper yarn — marshmallow, ginger, white, spice*
- ◆ *Tissue paper*
- ◆ *White paper*
- ◆ *White gummed tape*
- ◆ *Double-sided tape*
- ◆ *Cork pin board*
- ◆ *Tapestry needle*
- ◆ *Glass head pins*

COILED STRING FRAME

1 Cover the front of the frame with double-sided tape, making sure that the tape goes to the edges of the aperture and the frame.

2 Cut two long lengths of ginger flat paper yarn to twist together to form a thicker round string. Tie two cut ends to a door and tie the other ends to a pencil. Hold the paper yarn fairly taut and rotate the pencil with your fingers until the string is tightly twisted.

3 Cut the knots off the paper string and stick it to the inside edge of the frame aperture. Continue working around the

frame aperture, folding the corners neatly to create the appearance of lines radiating to the corners of the frame.

4 Once the front is covered with string, stick double-sided tape to the edge of the frame. Continue wrapping the string around the frame to cover the sides.

5 Trim the end of the string and tuck the end underneath the previous rows. Press the string firmly in place to stick.

WRAPPED TISSUE FRAME

1 Cover the front of the frame with double-sided tape in the same way as for the coiled string frame above. Cut thin strips of double-sided tape to fit under the rim of the aperture. Stick a single width of the tape to the back side of the frame.

2 Cut 1in/2.5cm strips of tissue paper. Take one strip and stick one end under the rim of the frame. Wrap the tissue around to stick to the back side of the frame and around onto the front and finish under the rim of the aperture.

3 Continue working all the strips from the back to cover the frame, allowing the tissue paper to crease attractively (*see picture next column*).

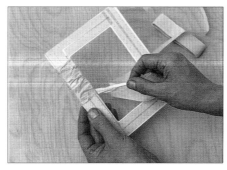

4 Wrap the tissue around the corners in the same way. Add extra double-sided tape under the rim if required.

5 Apply white gummed tape to the back side of the frame to cover the edges of the tissue paper and hold it in place.

WOVEN FRAME

1 Make a paper template of the front of the frame. Stick this to the pin board. Stick double-sided tape around ¾–1¼in/2–3cm from the outside edge of the template.

2 Cut lengths of paper yarn to fit lengthwise between the strips of tape (warp threads). Remove the backing from the top and bottom tapes and stick the lengths down leaving a small gap between each piece.

3 Weave lengths (weft threads) of yarn under and over the warp threads. Once you have woven several weft threads, remove some of the backing paper from the side double-sided tape and stick the paper yarn in place.

4 Continue weaving in the weft threads. Use the tapestry needle to ease the yarn into position. As you work down the frame, insert glass-headed pins to hold

the middle of the weft in position and then pull the ends out to straighten.

5 Once the template is covered, ease the weaving off the tape. Follow Step 1 of the coiled string frame and stick the woven frame in position.

6 Cut across the weft threads between the warp threads near the edge of the aperture. Cut across the warp threads between the weft threads.

▶ 87

7 Stick a narrow strip of double-sided tape on the inside of the aperture rim. Unravel the weaving inside the aperture and stick the threads on the back side.

8 Stick double-sided tape on the back side of the frame. Wrap the threads around and stick down. Cover the ends with strips of white gummed paper.

woven baskets

✳

Choose neutral shades of paper string to complement the natural color of the corrugated cardboard.

MATERIALS

- ◆ *Single-ply/wall corrugated cardboard*
- ◆ *Paper string – 2 skeins of each – chocolate, curry, straw, ash, snow*
- ◆ *Two skeins of flat chocolate paper yarn*
- ◆ *Pencil*
- ◆ *Safety ruler*
- ◆ *Craft knife*
- ◆ *Cutting mat*
- ◆ *Tapestry needle*
- ◆ *Brown paper bag*

1 Enlarge the template and cut the shape out of the cardboard.

2 Measure and mark one of the large panels in the template to divide it into 10 equal sections. Measure ⅛in/3mm either side of each line for the slits.

3 Lay the template over the cardboard shape and mark the top and bottom of the slits with the point of the craft knife (*see picture next column*).

4 Score the cardboard shape to mark the base of the basket. Cut out the slits up to the score line.

5 Mark and cut the slits on the other large panel. Turn the template around and position it centrally on one of the end panels. Mark and cut the slits in the same way.

6 Beginning in one corner, weave a skein of chocolate brown string over and under the card strips. When you get back to the first corner, wrap the skein round the tail end and start weaving in the opposite direction (*see picture next column*).

7 As you weave, the sides of the basket will automatically stand up. Finish the skein in the corner where you began, leaving a tail on the inside.

8 Begin a second color, weaving around and back in the same way. Tie the two colors together in a secure square knot and trim the ends neatly.

9 Add in the different colors. Try to keep the weaving level, but if necessary this can be corrected once the weaving is complete.

10 Stop weaving about ⅜in/7mm from the top. Cut ⅝in/1.5cm wide strips from the paper bag. Fold the strip over the top edge of the basket.

11 Oversew the top edge enclosing the brown paper with the yarn and sew in the ends.

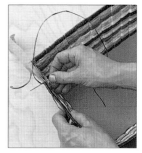

lace paper bowls

✳

The papier-mâché bowls can be made to different strengths. Do not use any solid areas in the paper to give the bowls a light, airy appearance.

MATERIALS
- *Japanese lace paper – mizutamashi and rakusui usumaki*
- *Large plastic bowl*
- *Vaseline*
- *PVA glue*
- *Paste brush*
- *Pressed flowers, leaves or feathers*

1 Smear the outside of the plastic bowl with Vaseline.

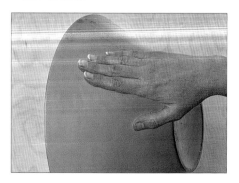

2 Tear pieces of Japanese paper into strips about 1in/2.5cm wide and then into ragged pieces about 3in/8cm long.

3 Dilute the PVA glue with five times as much water. Brush the paper strips with the glue and lay onto the bowl overlapping the edges slightly.

4 Cover the entire bowl with several layers of paper leaving the top edge very ragged. Leave to dry overnight.

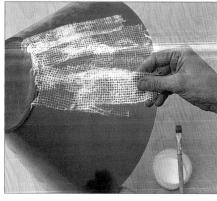

5 Squeeze the plastic bowl to ease the papier-mâché structure off the sides. If necessary you can slip a palette knife gently down between the bowl and the papier-mâché.

6 Remove the plastic bowl and test the strength of the papier-mâché. If necessary, replace the plastic bowl and add one of two more layers of the Japanese paper.

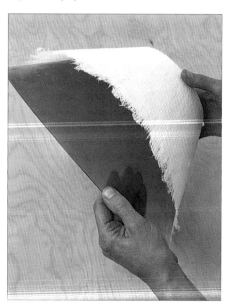

▶ 89

7 With the plastic bowl still in place, decorate the papier-mâché with pressed flowers, leaves or feathers if desired. Paste the additions onto the outside of the bowl and then tear pieces of Japanese paper to cover and secure them. Allow to dry completely again before removing the plastic bowl.

Japanese lamp

✳

It is possible to substitute brass rods for the thin/mild steel rods, which can be soldered together.

MATERIALS
- ◆ ¼in/5mm diameter thin/mild steel rods – four of each length of 22½in/56cm, 12¼in/31cm, and 12in/30cm
- ◆ Matte black spray paint
- ◆ Four sheets of poster-size/A2 Japanese shoji paper
- ◆ Noodle wrappers
- ◆ Japanese text
- ◆ Black ink pen
- ◆ Ruler
- ◆ Handmade Japanese paper
- ◆ White tissue paper
- ◆ Spray adhesive
- ◆ Double-sided tape
- ◆ All-purpose glue
- ◆ ⅛ in/3mm panel of MDF

1 The rods need to be welded together. Ask your local welding service/blacksmith (look in the Yellow Pages) to build the frame with 4in/10cm legs and 12in/30cm between the top and bottom bars. The shorter bars go at the top, so that the frame is wider at the base.

2 Spray the frame with black paint.

3 Enlarge Japanese text on a photocopier until the symbols are ⅝in/1.5cm high. Use the sample on page 105 or choose your own. Put the text under the shoji paper and trace directly onto the paper, drawing groups of symbols in a vertical line or square formation. You will need five or six groups for each panel of the lamp.

4 Hold a ruler on the back side of the paper and tear down the sides of each group of symbols to make square or rectangular panels. Tear carefully to leave a fluffy edge.

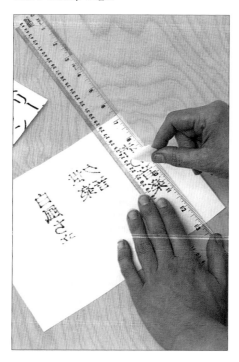

5 Draw a template for the panels of the lampshade. Lay a piece of shoji paper on top.

6 Cut the noodle wrappers into sections and arrange them within the lines of the template. Tuck the hand drawn text underneath the noodle wrapping. Once you are happy with the layout, remove all the pieces and spray them with adhesive, then stick them down in the same arrangement as before (*see picture next column*).

7 Tear pieces of handmade Japanese paper and stick overlapping the text and noodle wrapping. Then stick pieces of white tissue paper down in the same way to create areas with several layers of paper.

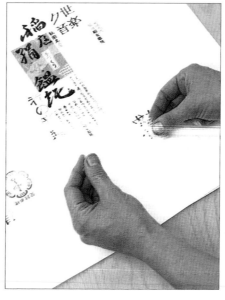

8 Trim the shoji paper allowing an extra ¼in/5mm down each side and ½in/1cm at the top and bottom.

9 Cut thin strips of double-sided tape to fit along the four bars of the first side of the frame. Position the shoji paper and press on to the frame.

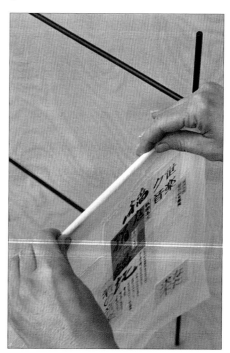

10 Trim diagonally across the corners and stick the allowance to the inside along the top and bottom edges.

11 Continue sticking the side panels one at a time to the lamp. Stick thin strips of double-sided tape down the side edges of the paper before attaching to the frame.

12 Cut a 4 x 12½in/10 x 31cm wide panel of MDF and glue to the base to support the light fitting.

concertina wall light

✳

Real gold leaf requires special equipment and considerable skill to apply. I have used "Dutch" gold leaf that is sold in little booklets. It is much easier to handle, but needs to be coated with shellac to keep it from tarnishing.

MATERIALS
- ◆ Semicircular wall light fitting
- ◆ Tobacco paper
- ◆ Parchment
- ◆ Stencil film
- ◆ Craft knife
- ◆ Ochre yellow stencil paint
- ◆ Stencil brush
- ◆ "Dutch" gold leaf
- ◆ PVA glue and Shellac
- ◆ Small piece of sponge
- ◆ Spray adhesive
- ◆ Paintbrushes
- ◆ Hole punch
- ◆ Fine tan leather thong and gold bead
- ◆ Newspaper
- ◆ White gummed tape

1 Trace the leaf template and outline in black. Place the stencil film over the motif and cut out along the lines with the craft knife.

2 Cut the tobacco paper and parchment in half lengthwise. Lay the tobacco paper on a flat surface. Using an almost dry brush, stencil the motif onto the paper. Move the stencil and repeat until there are leaf sprigs scattered all over the paper (see picture next column).

3 Clean and dry the stencil. Reposition the stencil overlapping one of the leaf motifs. Brush the PVA glue carefully over the stencil film and lift off.

4 Cut a sheet of gold leaf in half. Lay it carefully over the motif and press down carefully with a soft dry brush. If required, tear off excess gold leaf to patch other areas of the motif.

5 Allow to dry for a few minutes and then brush lightly with a soft brush to remove the excess gold leaf. Keep the scraps in a container to use at a later date. Complete all the leaf motifs in the same way.

6 Lay the paper face up on several layers of newspaper. Using a piece of sponge, wipe the shellac over the entire surface and allow to dry. Dispose of the sponge carefully.

7 Spray adhesive over the parchment and on the back side of the tobacco paper and stick together. Smooth the papers together with your hands or a roller to remove any air bubbles.

8 Trim the long sides of the panel to measure 12in/30cm wide. On the back side, mark every 1in/2.5cm. Use a blunt tool to score between the marks on the back side only.

9 Fold the paper along the score lines to form a concertina. With the right side

facing up, trim the panel so that the last fold at each end is a larger "mountain" fold, with the cut edge flat on the table.

10 Punch a small hole through the middle of each panel of the concertina, 1in/2.5cm in from each long edge.

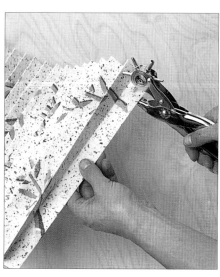

11 Thread the leather thong through. Lift the concertina onto the lamp and secure the paper to the back side with white gummed tape. Pull the leather thong taut and tape securely.

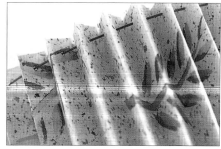

12 Replace the light switch cord (if there is one) with leather thong and attach a large gold metal bead to the end.

tucked and stitched lampshade

✳

Fixing ingredients such as salt and soda are only required if the dyed article is going to be washed. This project requires some sewing skills, carefully read through directions before proceeding.

MATERIALS
◆ *White drum lampshade 8¼in/21cm high, 11in/28cm in diameter*
◆ *Three sheets of Lx/lens tissue/tissue-tex paper*
◆ *Mint green procion "m" dye*
◆ *Sponge brush*
◆ *Sewing machine and equipment*
◆ *Double-sided tape*
◆ *Self adhesive white paper tape*

1 Mix a level teaspoon of dye powder with water in a jam jar to make a dye solution from a mint green cold water dye or use the procion "m" dye powder. If making it up yourself from powders, use 80 parts turquoise, 30 parts chrome yellow and 5 parts red.

2 Lay the Lx/lens/tissuetex paper on several layers of newspaper and brush with the dye solution. Lift onto fresh newspaper and allow to dry flat.

3 Cut each paper in half lengthwise. Mark every 1in/2.5cm down each long side.

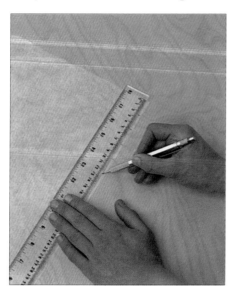

4 Fold the paper between the first two marks and flatten the crease with your finger. Using the side of the presser foot as a guide, machine stitch approximately 3/8in/8mm from the fold.

5 Fold the paper between the next two marks and stitch a second tuck in place. The fold of this tuck should touch the stitching of the first. Continue folding and stitching until the entire panel is complete.

6 Stitch five panels of tucks all together. With right sides together, stitch the panels together along the short edges maintaining an even distance between the tucks.

7 Trim the seam allowance to 1/8in/5mm and press with a cool iron. Press the tucks in the same direction.

8 Fold the panel in half lengthwise. Mark every 2in/5cm out from the center fold along the short edges of the panel. Fold the panel between the marks to make five lengthways creases in all.

9 Machine stitch down the central fold in the direction of the tucks. Stitch the two outer creases in the same direction.

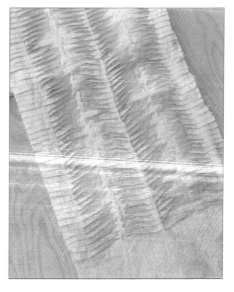

10 Turn the panel around and line up one of the other creases under the presser foot. Lift the first few tucks, fold over and stitch flat. Continue lifting a few tucks and stitching down in the opposite direction to form a zigzag effect with the tucks. Stitch down the remaining crease in the same direction.

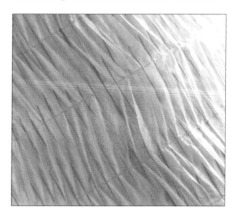

11 Trim the threads close to the paper. Stick thin strips of double-sided tape around the top and bottom rim of the lampshade and down the back seam. Trim the stitched panel close to the tucks at one end and stick centrally on to the back seam. Stretch the bottom edge slightly and stick the first inch. In order to keep the tucks vertical, ease the top edge and press onto the double-sided tape.

12 Continue fitting the stitched panel to the frame, stretching the bottom edge and easing in the top edge. Once complete, trim and fold under the end neatly covering the other raw edge.

13 Trim the top and bottom edge of the stitched panel to 1/2in/1cm. Fold to the inside of the lampshade and cover the raw edges neatly with white tape.

crocheted lamp base

✳

Considerable crocheting skills are required for this project, be sure to understand all directions before proceeding.

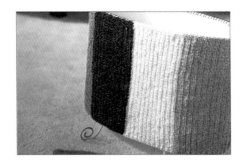

MATERIALS
♦ *Paper string – seven skeins each of indigo, lavender, lilac and white*
♦ *Size 2 crochet hook or suitable size hook*
♦ *Oval lampshade 8in/20cm high, 12in/30cm long*
♦ *Large needle*
♦ *Galvanized wire*
♦ *Parallel pliers*

1 Work a single chain with indigo paper string 8¼in/21cm long. This picture shows how to "yoh" by looping the string around the hook ready to pull through the chain.

2 Working back along the single chain, skip one chain and insert the hook under the top loop of the next chain. Yoh (see step 1), draw through the chain only. Yoh and draw through both loops on the hook. This forms one double crochet. Repeat to the end of the chain.

3 Turn and work a single chain, insert the hook through the second loop along and continue working double crochet as before. Keep the number of stitches constant so that the sides remain straight.

4 Use approximately three skeins of each color. Tie the next color in with a square knot and keep all the knots to one side. Crochet until the panel fits around the lampshade when stretched slightly. Finish off by pulling the end of the string through the loop on the hook.

5 Trim the knots neatly and wrap the crochet round the lampshade with the knots to the inside. Oversew the crochet along the top edge, using matching string. Tie in the next color so that the knot lies invisibly under the rim of the lampshade (*see picture next column*).

6 Oversew the back seam before attaching the crochet along the bottom edge.

7 Cut four 8in/20cm pieces of galvanized wire. Bend one end of each piece into a loose coil. Wrap the end of the coil with a thin strip of double-sided tape. Open out the end of the paper string and fold over the end of the wire so that it sticks to the tape. Wind the paper string tightly around the coil and halfway up the wire. Cover one wire in each of the four colors.

8 Attaching the feet depends on the inside fitments of the lamp. Twist the wire to hold the feet in position or tape securely to the inside.

p a r c h m e n t   l a m p s h a d e

✳

Add a touch of color to the soft tones of this simple design by using dyed natural raffia or adding beads between the bands of notches.

MATERIALS
- ◆ Natural parchment
- ◆ Coolie lampshade
- ◆ Lightweight white paper
- ◆ Natural raffia
- ◆ Craft knife
- ◆ Cutting mat
- ◆ All-purpose glue
- ◆ Double-sided tape
- ◆ Tapestry needle
- ◆ Black pen
- ◆ Pencil

1 Open out the old lampshade to draw a template on the white paper. Cut out and fold the template in half three times to make eight equal sections (*see picture next column*).

2 Open out and fold the last section only in half three times to subdivide it into eight. Mark down the fold lines every ½in/1cm with a pencil and join the dots to make curved guidelines.

3 Mark V-shapes with the pen down the third, fifth and seventh fold lines using the pencil lines to keep the V's straight.

4 Cut the parchment lampshade out using the template. To do this, place the template on a cutting mat and position the parchment on top.

5 Cut the V-shapes straight through the template. Open out each V-shape with the end of the craft knife. Move the next

section of the parchment round and cut another batch of V-shapes.

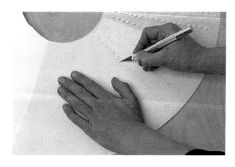

6 Continue in this way until each of the eight sections of the parchment has a band of V-shapes. If desired, cut a circle of V-shapes in between each band.

7 Overlap the back seam of the parchment, using double-sided tape to stick.

8 Spread a thin layer of glue round the rings of the lampshade and allow to dry until tacky. Fit the top ring first and then lower the shade onto the base ring.

9 Allow to dry, then oversew around the top and bottom edges using natural raffia. Sew in the ends neatly to complete.

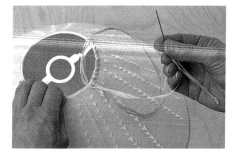

concertina blind

✳

The pale green rectangles in this design are made from Lx/lens tissue/tissuetex paper. This has wet strength and therefore can be dyed quite easily.

MATERIALS
- *Pale pink French bond paper/Mi-tiente cartridge paper*
- *Pale green Lx/lens tissue/tissuetex paper*
- *Spray adhesive*
- *Pencil*
- *Ruler*
- *Neoprene sheet*
- *All-purpose glue*
- *Burgundy red acrylic paint*
- *Small sponge roller*
- *Black 0.5mm drawing pen*
- *Hole punch*
- *Fine white cord and beads*

1 Cut long rectangles from the Lx/lens tissue/tissuetex paper. Spray the pieces with adhesive and stick them to the bond paper in a random arrangement (*see picture next column*).

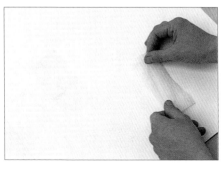

2 Draw a light pencil line lengthwise through each of the green rectangles at varying angles.

3 Cut the stamp shapes from the neoprene sheet. Glue each shape to a block of wood. Mark the exact position of the shape on the reverse side of each block.

4 Use a sponge roller to ink the stamp with paint. Line the stamp block up with the pencil lines and print two or three stamps along each one.

5 Draw in the line between the stamps with the pen. Add details such as the spiky flower head, large dot and small bar shapes (*see picture next column*).

6 Mark every 1½in/4cm down the sides on the back side of the blind. Score across the blind between the marks with a blunt tool.

7 Fold the blind along the score lines to form a concertina. Join two pieces of concertina together by overlapping the paper so that the cut edge tucks into the fold line. Glue in position.

8 Close the concertina and punch a hole 1¼–1½in/3–4cm from the side edges.

9 Thread the fine white cord through the holes. Fit attractive beads at the bottom of the blind. See page 106 for details on fixing the blind to the window frame.

window box

✽

Khadi paper has a soft texture that can be cut easily but is stiff enough to hold its shape.

MATERIALS

- *22oz/640g weight khadi paper*
- *Carbon paper*
- *Flat acrylic milk-white paint/Reproduction milk paint*
- *⅛in/3mm thick MDF cut to height of 5in/12cm x width of window*
- *⅝in/1.5cm thick wood and screws*
- *Wood glue*
- *Pencil*
- *Craft knife and cutting mat*
- *Safety ruler and scissors*
- *Tape measure*
- *Round file*

1 Transfer the template to the khadi paper using the carbon paper (*see picture next column*).

2 Measure the width of the window frame with the tape measure and move the template along the khadi paper marking out the whole of the border to the required width.

3 Cut out all the petal shapes with a craft knife on the cutting mat.

4 Cut around the curved edges with scissors. Measure and mark the height of the window box. My screen is 8¾in/22cm high. Cut out with a craft knife and safety ruler.

5 Paint the screen and allow to dry. File any rough edges around the dentil work and then give the screen a second coat of paint.

6 Screw a 4¾ x ⅝in/12 x 1.5cm strip of wood to each front edge of the window recess. Glue the MDF panel in position. Stick the paper screen to the MDF.

shelf border

✳

Bockingford paper is a top quality artist's watercolor paper. Choose subtle tinted shades to enhance the design of this attractive border.

MATERIALS
- *Tinted Bockingford paper – cream, gray*
- *Scissors*
- *Hole punch*
- *Craft knife*
- *Cutting mat*
- *Pencil*
- *Tape measure*
- *Small paper fasteners*

1 Transfer the border template to a spare piece of Bockingford paper and cut out.

2 Measure the length of the shelf. Draw around the template, moving it along the edge of the paper and marking it out to the required length. Cut along the curved edge with scissors.

3 Punch the holes out of the template. Mark the position of the holes around the curved edges of the border. Move the template along the top edge to mark evenly spaced holes.

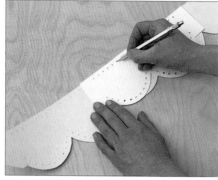

4 Punch the holes with a hole punch.

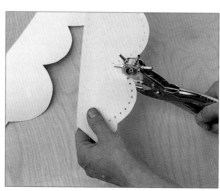

5 Cut a template of the bird, wing and flower pieces from spare Bockingford paper. Punch the holes in the templates. Draw around the large flower in the opposite color to the border and cut out. Cut the flower centers from the border color (*see picture next column*).

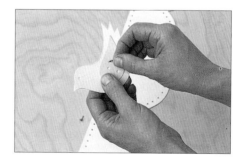

6 Cut the birds out of the opposite color to the border. Cut half with the template facing one way, then turn it over and cut the rest. Cut the wings from the border color. Punch holes in the wings and the bird.

7 Hold the wing in position on each bird and push fasteners through the holes.

8 Position the birds on the border above the large curves. Push the fasteners through the border and open out on the reverse side. Fix the flowers in the same way above the small curves.

9 Attach the border to the edge of the shelf with glue or double-sided tape.

leaf blind

※

Seek out textured handmade papers with fragments of grasses and petals that add texture to the leaf shapes.

MATERIALS
- *Two sheets of Lx/lens tissue/tissue-tex paper*
- *Orange and green handmade plant or flower papers*
- *Pencil*
- *Iron*
- *Craft knife and cutting mat*
- *Spray adhesive and glue stick*
- *Paper string and needle*
- *Curtain rod*

1 Trace the leaf templates and use them to draw out about 20 different shapes on each color of handmade paper.

2 Cut around the outline with the craft knife and then carefully cut out and remove all the small triangles representing the veins of each leaf.

3 Leaving a border of at least 2in/5cm around the edge, arrange the leaves on one sheet of Lx/lens/tissuetex paper until you have a pleasing and balanced pattern.

4 Fix the leaf shapes in position with spray adhesive.

5 Spray the entire sheet of Lx/lens/tis-suetex paper with adhesive. Match the bottom corners of the second sheet of Lx and smooth the paper towards the top with the flat of your hand.

6 Press under a ½in/1cm double hem around the four sides. Miter each corner for a neat finish. Glue the hem in place with glue stick.

7 Hang the leaf blind by making loops of paper string through the top with a needle and slipping it on the curtain rod.

tulip screen

✳

For a balanced appearance, make the bottom panels 4in/10cm taller than those on the top two rows and keep the same border around the panels except at the bottom. Make that slightly taller.

102 ◀

MATERIALS

- ◆ *Sugar cane paper in a range of colors*
- ◆ *Tracing paper*
- ◆ *Hard and soft pencils*
- ◆ *Handmade paper*
- ◆ *Glue stick*
- ◆ *PVA glue*
- ◆ *Three-fold screen*
- ◆ *Paint*
- ◆ *Spray matte varnish*

1 Enlarge one of the tulip templates, place a piece of tracing paper over it and draw over the outline with a pencil.

2 Transfer the shape onto the colored sugar cane paper for the main tulip.

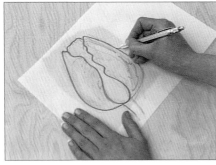

3 Carefully tear along the pencil lines. To ensure that the torn edge remains neat and accurate, work with the first finger and thumb of both hands. Keep your hands close together and tear a short distance at a time.

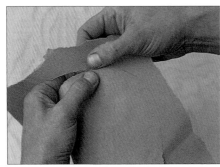

4 Following the main photograph tear out leaf shapes and a stalk in green paper. Stick the leaves and stalk in place on one of the background panels.

5 Stick the main tulip piece at the top of the stalk. Transfer the outlines of the

petal shading to different colors of sugar cane paper. Begin with the largest pieces and build up the layers until the smallest details are glued on top.

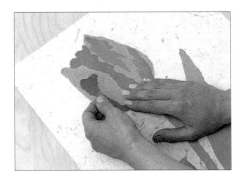

6 Complete nine panels in all, using three different background colors.

7 Paint the screen in a coordinating color. Leave to dry. Dilute the PVA glue 50/50 with water and use to glue the panels to the screen.

8 Spray the screen with matte varnish to protect the surface.

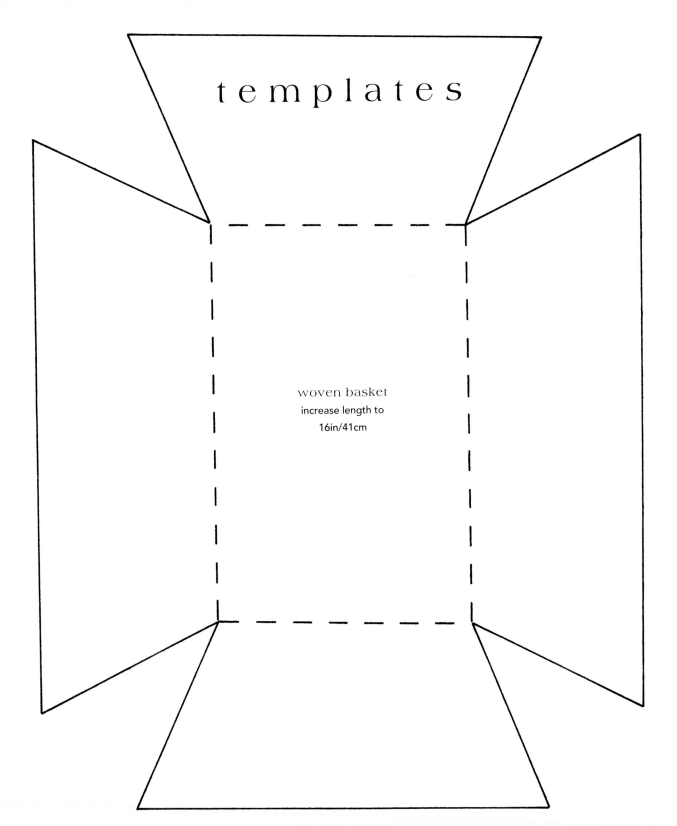

templates

woven basket
increase length to
16in/41cm

shoe box
increase length to
9¾in/24.5cm

黒人音楽をして以来、欲求不満のなった。べ物質的な豊

Japanese text
Enlarge to the size of
your choice

▶ 105

wall light
actual size

TEMPLATES

leaf blind
actual size

concertina blind
actual size

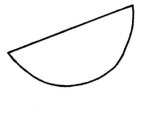

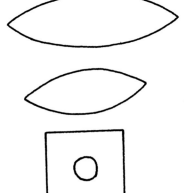

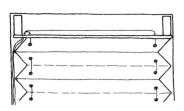

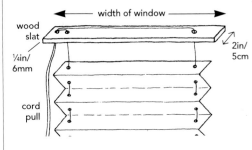

ATTACHING THE CONCERTINA BLIND

width of window

wood slat

¼in/ 6mm

cord pull

2in/ 5cm

wood blocks
½ x ¾ x 2in/1 x 2 x 5cm

1 Drill holes in a wood slat. Feed the cord through.

2 Screw two wood blocks into the window recess.

¼ x 2in/6 x 50mm MDF

cleat

3 Glue the top of the blind to the wood slat. Screw the slat onto the base of the small blocks.

4 Glue a piece of MDF to the front of the blocks with wood glue as a facing board. Secure with panel pins. Screw a cleat to the wall.

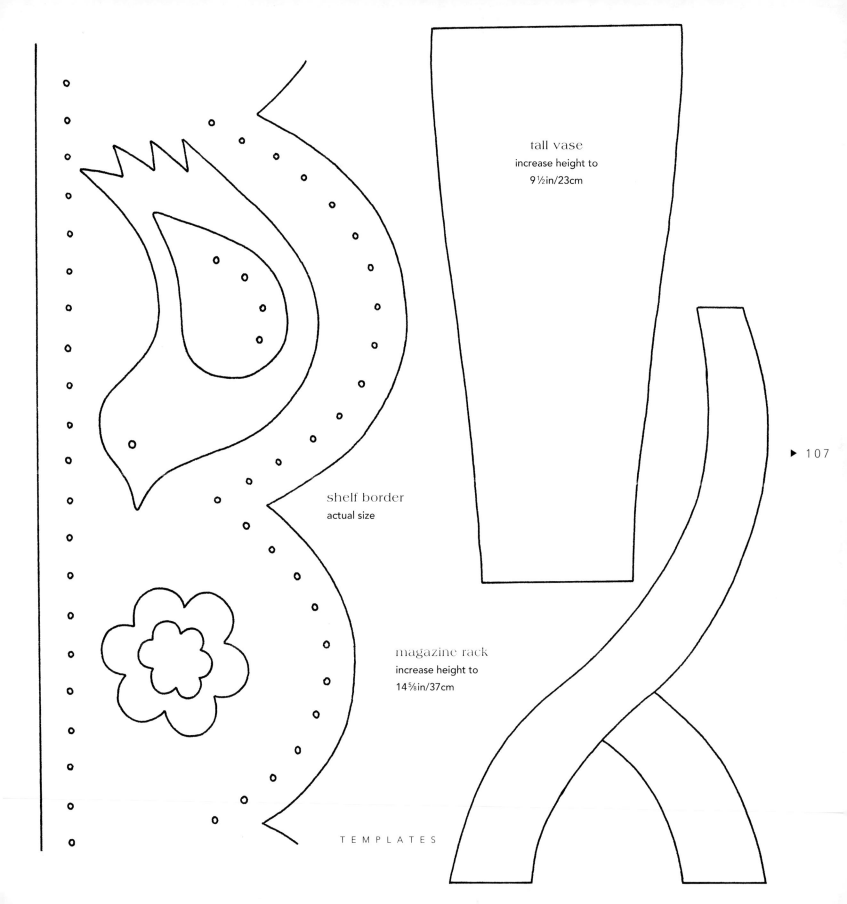

tall vase
increase height to
9 ½in/23cm

shelf border
actual size

magazine rack
increase height to
14 ⅝in/37cm

▶ 107

window box
actual size

tulip screen (1)
increase between
the markings to
10½in/26.5cm

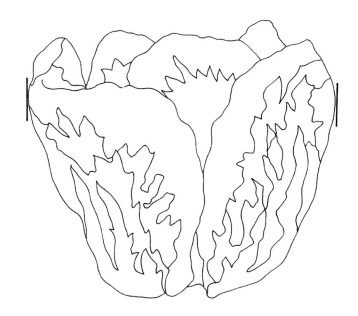

tulip screen (2)
increase between the markings to 6 1/4in/16cm

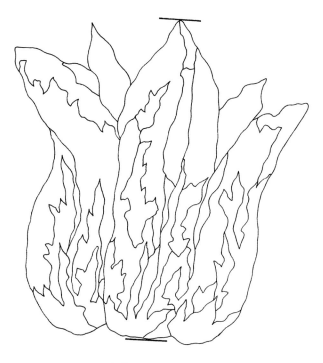

tulip screen (3)
increase between the markings to 6 1/4in/16cm

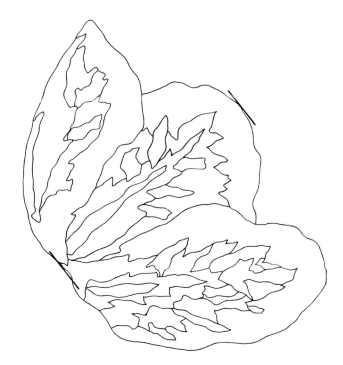

tulip screen (4)
increase between the markings to 5 7/8in/12.3cm

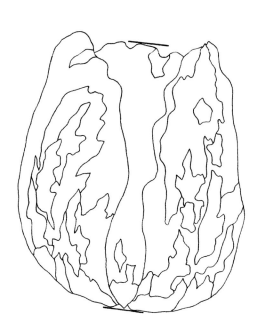

tulip screen (5)
increase between the markings to 5 1/4in/13.4cm

tulip screen (6)
**increase between the
markings to 8 ½in/21.5cm**

tulip screen (7)
**increase between the
markings to 8 ¼in/20.7cm**

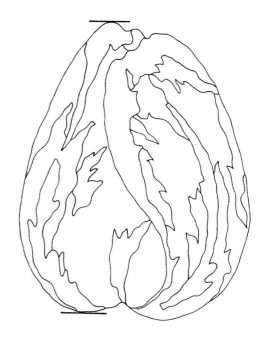

tulip screen (8)
increase between the markings
to 5 3/4in/14.5cm

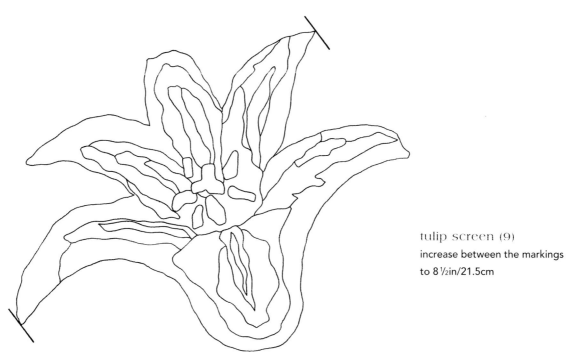

tulip screen (9)
increase between the markings
to 8 1/2in/21.5cm

index

✽